# WAS THE UNIVERSE TRULY CREATED BY GOD?

ALFONSO CASTILLO G.

Copyright © 2018

Alfonso Castillo G.

All rights reserved.

## INTRODUCTION:

Being a Philosophy student, still young and meeting frequently throughout the study of different subjects with the apparent contradictions between Positive Science and Reason; listening to the objections that the positivists make of the traditional method of Philosophy and also with the rejection of some philosophers and rigid thinkers by the conclusions of the new sciences, confronting each other throughout history, by not agreeing with such situation , because I always had intuited according to the logic that the truth is one and that one truth must shine in all planes, not leaving place to the contradiction, I decided then to penetrate also in the field of sciences; Physics, Chemistry, Astronomy, to know their essential postulates, their development and especially their conclusions, but there was still missing a very important aspect; the confrontation with the affirmations of the Faith emanating from the so-called Holy Books of the Bible.

Now, being an elder one and feeling myself in an advanced age, I have the purpose of finding the truth and this purpose is totally sincere and honest, without any assumption, with the firm intention of accepting the truth that I was looking for, whatever it may be.

In the same way, I invite the reader to assume a similar attitude in the reading and development of the present work.

The author,

Alfonso Castillo G.

## CONTENTS

|   | INTRODUCTION | i |
|---|---|---|
| 1 | Chapter One. WHAT DOES SCIENCE SAY? | Pg.# 1. |
| 2 | Chapter Two. WHAT DOES PHILOSOPHY SAY? | Pg # 18. |
| 3 | Chapter Three. WHAT DOES THE BIBLE SAY? | Pg # 33 |
| 4 | SUMMARY. THE ORIGIN OF THE UNIVERSE. | Pg # 42 |
| 5 | EXPLANATORY SHEET. | Pg # 50 |
| 6 | BIBLIOGRAPHY. | Pg # 52 |
| 7 | ABOUT THE AUTHOR. | Pg # 54 |

# 1.- CHAPTER ONE.
# WHAT DOES SCIENCE SAY?

Let's start by understanding what the word SCIENCE designate.
SCIENCE = From Latin SCIENTIA from SCIRE which means "TO KNOW." Let's resort to some definitions:
Science is the true knowledge that the human being is acquiring to discover the constants in the behavior of nature and after translating them into the rigor of mathematical formulations and empirical verifications, they are constituted in laws, as a consequence of such a systematic process.
*Science is all knowledge that can be demonstrated; rationally or experimentally.* In the case of rational proof, reducing the propositions to the evidence of the Four Logical Principles.

Briefly; *Science is knowledge by causes*. Generally, in the case of Natural Sciences, these causes are close and in the case of Rational Science, they are remote causes.

The fact that Science, through time in its evolution and overcoming of the known, seems to not always agree with the anterior knowledge, is attributable rather to the interpretation of the data than to the objectivity of it (no place for skepticism), which, of course, is perfectible by knowing day by day, more and more, that is, as soon as we have more elements, our perspective of vision is extended and so on.

From the ancient conception of the Earth as the center of the universe (*until then known*) with all the stars around, later supported by the **Ptolemy System** (Claudius Ptolemy 100-170 Greco-Egyptian astronomer, who exposes his ideas in his work "The Almagest") until the Renaissance when a new theory arose and that theory would change forever that conception; that of **Nicolaus Copernicus** (Polish Astronomer 1473-1543) in which it is no longer the Earth, but the Sun that is in the center with the planets orbiting in circles, man has always had the concern to know the origin and foundation of the universe, so, the first Scientific Revolution in experimental sense, begins here, in the sixteenth century, although at first, there were resistances for accepting this new vision, Copernicus himself refused to publish his writings until shortly before his death, finally, this new way of conceiving reality ended up to prevail.

Another great step in the advance of celestial science is given by Johannes Kepler (German astronomer 1571-1630) which specifies in three mathematical laws the exact movement of the planets. He tells us that the orbits of the planets are elliptical and not round, it is fair to mention also the name of Tycho Brahe (Danish astronomer 1546-1601) who contributes to the achievements of Kepler by providing him, in the communication between both, his vast experience in data observations made in sheets that he made throughout his life.

**Galileo Galilei (1564-1642)** scientist born in Florence, Italy; He tried to explain what was discovered before him. He defended the Copernicus system highlighting his contributions on the Principle of Inertia and the Development of the Theory on the Fall of the Bodies. By already having the first astronomical telescope whose invention some people attribute to him, he achieves notable advances in the knowledge of the movement of the stars.

In spite of his polemical and controversial personality, because due to many misunderstandings, he was confronted and even called to trial by the Inquisition, it can be said that he is one of the greatest constructors of the Scientific Method in the modern sense.

**Isaac Newton (1642-1727)** English scientist who culminates and unifies all the discoveries before him in the matter of celestial mechanics, perfects The Calculation in Mathematics, discovers the laws that govern optics, but his most important achievement is the Law of Universal Gravitation:

*"Every mass exerts on another an attraction of power proportional to their respective masses and inversely proportional to the square of the distance separating them."*

Pleasing even the most demanding ones of his time, Newton is undoubtedly one of the brightest minds in the History of science. It is thought then that all we had to know about the movements of the stars in space was already discovered and consolidated.

Albert Einstein (1879-1955) German scientist, it could be said that he initiates a second Scientific Revolution with his theories about Relativity based on the speed of light; *The speed of light is the same in all directions for all who observe it, whether they are in a state of rest or movement.* This, that seems not to be clearly understood, because it contradicts common sense, it's because light, not having mass, is not subject to the laws of inertia, thus they are photons, particles of pure energy.

The speed of light that is approximately <u>300,000 km per second</u> is always constant, regardless of the observer's position. Einstein also reveals that this is the highest speed at which something can travel in our reality.

Another of the contributions of this Theory (*expressed as Restricted and another later as General*) is the new concept of Gravity that occurs when space-time is curved by the bodies that possess mass. In the Theory of Relativity, emanating from the <u>constant</u> of the speed of light, space and time become relative. With Newton we believed that Gravity was a force, however Einstein's Theories were finally demostrated with observational data during solar eclipses, but the Theory of Relativity is much more and here comes what matters to us; The equations of Relativity formulated by Einstein that originally have the constant speed of light, describe the behavior of the universe. Although, at first Einstein thought that the universe was static and without beginning, in spite of that a Russian mathematician, specialized in Relativistic Cosmology; **Alexander Friedman** (1888-1925) makes him see that his equations describe a universe in expansion, Einstein despises him and even modifies those equations to support his point of view, however, as we know, later in his encounter with **Edwin Hubble** (American Astronomer 1889-1953), recognizes his error and accepts the beginning and the expansion.

We wanted to make this brief summary of the development of science from antiquity to the current times, aimed at making the reader see how the culture of knowledge has gradually come, especially in Physics and Astronomy, that is Astrophysics, the formidable discoveries demonstrated through the marvelous devices launched into space, such as the satellites **COBE** (*Cosmic Background Explorer*), **W-MAP** (*by its creator the scientist Wilkinson, to prove the fluctuations of the microwaves of the background radiation*) and **PLANCK** (*In honor of the scientist Max Planck, initiator of Quantum Physics*) which corroborate in an increasingly clear and irrefutable way the birth, formation and expansion of the universe.

Science tells us that everything started in a so-called BIG BANG or Great Explosion where they initiate, in unison, the ENERGY-MATTER and the SPACE-TIME, what we should understand, according to the greatest astrophysicists and theoreticians of this discipline is that THERE WAS NOT A PREVIOUS SPACE where such an event would take place, not even a real "BEFORE", because just there begins the STARTING OF TIME, so that, from the first moment of its existence, space-time began to expand more and more as this universe in its primitive beginning, was a great flash of pure energy, which, as it expanded, was gradually cooling and thus "condensing" in matter (*atoms, mainly Hydrogen and Helium that, by atomic reactions they would form the entire chain of the elements and would have to be converted, in the next millennia, always in expansion, into nebulas, galaxies and stars as we know them now*).

This Great Explosion, with which everyone agrees, should have happened approximately 15,000 million years ago (*Data of public domain verifiable in any book of Basic Astronomy as: "Un Universo en Expansión" by Luis F. Rodríguez, Fondo de Cultura Económica.*) that is, what a child would say is that human science has proven that everything started because "The Nothing" exploded.

## Is it possible that all this had happened spontaneously or randomly?

Some theorists of Quantum Mechanics affirm that it is possible, then according to them, at subatomic particle levels fluctuations occur, where in **an empty space of bodies** but **full of energy** the particles of matter spontaneously "jump" and then they make us an analogy that could be similar to the origin of the being of the universe, but this is a confusion, or rather a fallacy (*seemingly true, but false argument*) because that *empty space of bodies* that they comment, together with the energy are formally REAL BEING and our universe, as we have already said and we know, was not born from any "previous" but from "nothing" itself.

With the Theory of Relativity, also in Quantum Physics, the dualism between these two traditional substances of Classical Physics (Energy-body) has ceased.

Recall also that since Albert Einstein, we have been able to understand that energy and matter are reducible, that is, the real BEING has two phases or "faces" that is, it is presented as <u>waves</u> or <u>as solid corpuscles</u>, we reiterate then, that the quantum fluctuations are given in <u>what already exists</u> and do not come from the "nothing".

Positive science cannot go any further, because its field is the tangible, but is this enough? Can we settle and stay there?   OF COURSE NOT!!

*"The cause of all science, reduce fundamentally to the need to know, to understand as much as possible in length and depth the "how" and the "why" of things to reduce everything to a unity. This need is an essential feature of the rational being that we are; in it is expressed the greatness that consists in aspiring to be equated by thought to the totality of being." (Origin and End of Science, Logic and Cosmology, Page 160. Regis Jolivet, Ediciones Carlos Lohlé)*

Established, then, that the universe began in a Big Bang, we will see roughly how science came, step by step, to this formidable conclusion.

From the Belgian Astronomer George Lamaitre in the 20's (*20th century*) who, according to his mathematical calculations, tells us that the universe should be expanding. Later, the Astrophysicist George Gamow, around the year 1946, speaks clearly of a Great Explosion to explain the origin of the universe and its expansion. He also affirms; that if this explosion did not occur in a previous space, the great echo of it, should still "be heard" within the space that is "created" with the expansion.

This, then hypothesis, was finally demonstrated becoming the Great Theory of the BIG BANG (*Theory = Set of laws already demonstrated) (Introduction to the Scientific Method, Raúl Gutiérrez Sáenz, Editorial Sphinx, page 205*).

But how was this? The North American Astronomer Edwin Hubble, demonstrated through the redshift (*based on the Doppler effect*) that the galaxies present the universal expansion, elaborating his famous Hubble Law (*"The redshift of a galaxy is proportional to the distance it is"*) Hubble discovered that galaxies are moving away from each other, with a speed proportional to their separation. His observations through a large telescope allowed him to reach such a conclusion, because these galaxies, due to its displacement, had a small red tail.

Later, two engineers from the Bell Company, Arno Penzias and Robert Wilson, who were perfecting satellite communications, discovered The Background Radiation (*The echo of the Great Explosion that in the 40's Gamow had predicted*). At first they got confused thinking that it was a simple interference, but after an exhaustive review of their devices and with the intervention of the Physicist Robert Dicke, who was already looking for The Background Radiation, it was finally interpreted as the possible proof of the Great Explosion , but it is not until the period between 1989 and 1992, with the experiments carried out by the COBE (*Cosmic Background Explorer, an American satellite*) when this beginning of the universe officially remains as scientific proof. Arno Penzias and Robert Wilson were awarded the Nobel Prize for this discovery.

One of the first conclusions that emerge after establishing the beginning of the universe in the conditions already mentioned, in which today, all scientists agree, is that the matter is not eternal as was believed especially in the nineteenth century and until the middle of the twentieth century, then the bases of any materialist philosophical doctrine fall by themselves.

The Principle of the Conservation of Matter, in which someones believed to see the eternity of it, refers to the matter understood as *weight* and is born from the experiments of **Lavoisier** (*French Scientist 1743-1794*) who in 1772, found that in all reactions, including combustion, *the weight of the compounds is equal to the weight of the components before and after the reaction*, but time would let us know that things were not as simple as they were then believed, because the most surprising of the reactions was unknown; the Nuclear Reaction. In this, under the concept of matter understood as <u>*weight or mass*</u> it is impossible to understand what was previously concluded, due to the so-called "Mass Defect"; *that little radiant energy that is lost and leaks in the form of light when atomic fusion takes place*.

<u>Foundation of the above;</u> *Our Sun, like all stars, is a sphere of gases in atomic reaction that radiates energy in this way; "The Sun converts some 654,600,000 tons of Hydrogen into something less than 650,000,000 tons of Helium per second. It loses therefore 4 million 600 thousand tons of mass every second, but even at that tremendous rate, the Sun contains*

enough Hydrogen to remain active for billions of years more. *"(Check in: "A Hundred Basic Questions About Science" by Isaac Asimov, Alianza Editorial, Pages 50 and 51)*

This light, like that of the Sun (*which is an atomic furnace like all stars*) wanders through space and if it is absorbed by an opaque body, it is transformed into heat and it carries out work, movement and / or cools, so, concentration of the energy is disappearing very slowly but necessarily, this because the universe EXPANDS !! Energy and heat "is not enough" to keep it stable.

Then, even the famous sentence of the past "*nothing is lost, nothing is created, everything is transformed*" remains in question, stumbling   WHY?       Well ...

BECAUSE WE HAVE SEEN THAT EVERYTHING BEGAN !! AND IT WILL END AS SCIENCE HAS ALSO PREDICTED, IN A COLD DEATH, then, it turns out that the verifiable and constant transformations that we see in nature, only occur in the transit of the above described while the universe lasts.

The Cold Death of the universe, that is, the cessation of any combination and translation of the hot to the cold with the respective enormous degradation of energy, because what remains of the universe, will expand indefinitely, <u>since the universe is an open system</u> .

The Boomerang Project of NASA had opened the possibility of an unbridled expansion and in recent years, with the great Hubble telescope the existence of dark energy has been proven, that still not understood repulsive force that is 73% of the universe "shoots" the furthest galaxies, increasing their speed. (*To corroborate this, the reader must go to the latest sources that science has discovered for this particular point.* (https://ciencia.nasa.gov/science-at-nasa/2001/ast03apr_1))

Cutting off any **imagined Big Crunch** (*contraction of the universe at a critical point, due to its mass by gravity, which according to some, would happen repetitively*) and leaving us only the **certainty of a disintegration to infinity**, as up to now everything points at.

But as we had said before, the need to know more about our material space-time reality, cannot stop in this and certain Big Bang. We cannot stay there, because the legitimate desire to know and understand is not satisfied only with the "WHAT?" And "HOW?" You also need the "WHY?"

Then, it is up to **reason** to give us the continuity. The data obtained in the scientific experiences (*positive science*) must be reasoned and inserted in the global context of knowledge to have its real intellection, **this is the true science, which is all knowledge that can be demonstrated.**

### EXPLANATION:
"*The fact of scientific proof does not rely exclusively on the input of the senses, but it is possible to establish a foundation of an intellectual type. So it is feasible to call science to Mathematics, Physics, etc. as to Philosophy. The key idea is to discard "verification" as synonymous with "empirical verification" ". (Introduction to the Scientific Method of Raúl Gutiérrez Záens. Page 10. Editorial Esfinge*)

Throughout the History of Philosophy, there have been thinkers who, distrusting reason, have denied Metaphysics (*Metaphysics = the things essentially inexperienced, immutable and in some spiritual way, but which can be discovered by reason. Philosophy Dictionary. Walter Brugger, Herder Library*) especially **Immanuel Kant** (*German Philosopher 1724-1804*) because he tried to bring the Physico-Mathematical method ... to Philosophy !!!

Kant, dazzled by Isaac Newton's recent discoveries and achievements; The Celestial Mechanics, especially The Law of Universal Gravitation (already described above) which marveled everyone in his time, because men at that time were sure they finally knew how the universe worked.

Later, as we will explain afterwards, Einstein's Theory of Relativity would question and overcome some of the Newtonian conclusions, but let's continue with Kant; by distrusting reason, also influenced by **David Hume**, English empiricist (1711-1776) (*Empiricism = only what can be perceived by the senses is valid for knowledge*) in his book "Critique of Pure Reason" makes a serious mistake by denying the validity of Metaphysics as a science.

What to say?

OF COURSE!! Metaphysics is not a POSITIVE science but YES, IT IS A RATIONAL SCIENCE !! As valid as Physics or Mathematics!!

Also the same positive science shows the evidence of the falsity of Empiricism, in which Hume dares to deny the Principle of Causality because according to him, we have no impression of the cause as such it is.

As we already know, Hume influenced Kant so profoundly that he dares to say that Hume woke him up from his "dogmatic dream". The influence of Hume is also strongly felt in **Bertrand Russell** with the erroneous consequences that we will later corroborate.

## Brief demonstration of the falsity of Empiricism:

*Hume said that we would never know what the stars are made of, because we cannot touch them and that we could never affirm the roundness of the Earth with arguments of reason by not having a complete impression of it; but again the same positive science would put in evidence this erroneous way of thinking, because in Astronomy, thanks to a modern research apparatus; The spectrograph which decomposes light to investigate in its spectrum what kind of element comes from, because each atom has a specific spectrum, today we know with certainty what the stars are made of and thanks to the space missions, to mention some, NASA's Apollo Project on its return from the Moon, we have a beautiful photograph of Earth known as "The Blue Planet."*

*Is it that until that moment the Earth was round? Or rather, spherical? OF COURSE NOT!! then, the rational deductions of the past, which already affirmed it, were right! Now that we understand all this, we can say: Hume's Empiricism, rest in peace. Amen."*

As previously explained; also Reason is able to demonstrate the arguments overcoming to the 4 Logical Principles:

*(Principle in Logic = Proposition that does not need proof because it is evident and universal and serves as the basis for any demonstration)*

1.- **_PRINCIPLE OF CONTRADICTION_**. It can be stated as follows: <u>TWO CONTRADICTORY PROPOSITIONS CANNOT BE BOTH TRUE</u> "It is impossible to BE and NOT TO BE at the same time."

2.- **_PRINCIPLE OF IDENTITY_**. It is stated as follows: "A" IS NECESSARILY "A". That is, every being is identical to itself in a moment and under the same aspect.

3.- **_PRINCIPLE OF THIRD EXCLUDED_**. THERE IS NO "MIDDLE TERM" AMONG TWO CONTRADICTORY PROPOSITIONS, that is, <u>half-truths do not exist</u>.

4.- **_PRINCIPLE OF CAUSALITY_**. EVERY CONTINGENT BEING HAS A CAUSE, or every effect requires a proportional and sufficient cause for the qualities of that effect. We could also say, to better understand it, in a negative way: "NOBODY GIVES WHAT DOES NOT HAVE."

Therefore, <u>EVERY PROPOSITION OR KNOWLEDGE THAT CAN BE DEMONSTRATED IS SCIENTIFIC KNOWLEDGE</u>.

The same principles of Classical Mechanics have their corresponding endorsement in the Reason, for example; action-reaction is only a positive statement of the principle of causality (*each impulse corresponds to a proportional reaction being this strictly necessary*).

It should also be highlighted here the error and arbitrariness of the pure positivists (*Positivism = The exaggeration of only having experimental science as the sole criterion of truth*) not wanting to recognize the value of Reason to declare the conclusions obtained by it as "lack of sense", especially the English mathematician and philosopher Bertrand Russell (1872-1970) who developed a whole doctrine of disquisitions based only on mathematics.

## MY EXPLANATION:

If one morning, I'd discovered myself in a room, and after becoming aware of it and investigating a little about it, I'd discover that it is part of a house and after getting to know it and understanding the practical disposition of the rooms, also its orientation and even the probable time from its construction until now and I'd even come to know exactly the intimacy of the nature of the materials with which it is made of and the principles that govern it so that it remains standing up (*up to here the positive science*), we recognize, all this knowledge is certain, valid and indispensable but in no way "lacks sense" and it is also essential to ask myself; WHO AM I ?, WHAT AM I DOING HERE ?, HOW DID I GET INTO THIS SITUATION?, WHO PLANNED AND BUILT THE HOUSE?, WHAT ARE MY SPECTATIVES and then WHAT MY DESTINY WILL BE? (*This is the task of Philosophy called until the nineteenth and twentieth centuries as Metaphysics*), because mankind must know its origin and identity in reality to decide its behavior in order to its final destination.

Needless to say that this previous figure of the house is humanity in the universe.

I reiterate, only recognizing the certain and true conclusions of positive science; the "WHAT?" and "HOW?" and ignore the also certain and true ones of Philosophy; The "WHY?" of the ultimate causes, leaves us with only part of the knowledge when we must aspire to the whole of it.

**ALTERNATE PROOFS OF THE BEGINNING OF THE UNIVERSE.**

(A) As one of the consequences of the Expansion of the Universe, to put it in the simplest way possible, we know; **THINGS COOL UP.**

**EXPLANATION:**
Our Planet Earth began as an incandescent mass that step by step was cooled down to be able, on its surface, to house life, as we all know. However, its core is still incandescent as we can see from volcanic eruptions.
If the universe had been cooling from eternity, it would already be totally cold and dead, but it is not, there are still very hot stars and however they are cooling little by little, therefore this process started in a concrete past.

(B) Hydrogen, the first and most abundant element in the universe, is not renewable. In the Big Bang it was produced in large quantity along with Helium. The Hydrogen, when entering in atomic reaction becomes Helium and this one, in entering in atomic reaction, forms another element, until completing, in a successive chain, all the elements known until now.

But the important thing of this is that if this transformation had been done from eternity, there would be no hydrogen in the universe. So there is abundant hydrogen still today, then, the universe began at a specific moment in the past.

These two tests help us have a more perfect and total certainty of the non-eternity of matter.

# CHAPTER 2.
# WHAT DOES PHILOSOPHY SAY?

Let's start by explaining what Philosophy is to have a notion of it. The nominal definition is very romantic; **_Love for wisdom._**

According to an old tradition, Pythagoras (*570-497 BC*) being praised and called wise, replied modestly: "*I am not wise, only a lover of knowledge*" (*Philosopher*) (*www.acfilosofia.org*).

Conceptually, Philosophy is the knowledge for remote causes that reaches the reason in its search for the truth until the last why on topics such as; the BEING, the origin of the world, good and evil, the final destiny, etc.

When and where is Philosophy born?

In ancient Greece, when man chooses to investigate for himself and using his reason, he decides to break with the myth; the traditions and beliefs that passed from generations to generations as an explanation of reality (*Mythologies*).

It is here, then, where scientific development begins. The man observes, reasons and concludes. However, we must say that the first conclusions were not always the most fortunate, but, nevertheless, in that attitude, science was born.

The first Philosophers are called "The Pre-Socratic" (Before Socrates, Greek Moralist Philosopher 470 B.C. - 399 B.C.) and are: Tales of Miletus, Anaximander and Anaximenes, (seventh century B.C.)

Philosophy, initially, encompassed all knowledge, all disciplines, all subjects, but gradually, in the course of History, to define each specialty, that is; a science that studied the numbers, another the life, another the strictly material, etc. these were separated until they are as we know them today.

Since the origin of Philosophy, there have been many trends and ways of conceiving reality, but we can summarize them all in THREE; REALISM, IDEALISM AND MATERIALISM.

REALISM: ACCEPTS THE EXISTENCE OF SPIRIT AND MATTER. Main representatives: Aristotle and Thomas Aquinas. Ancient Greece and Scholastic respectively. Jacques Maritain, French philosopher (1882-1973), Étienne Gilson, French philosopher (1884-1978), Frederick C. Copleston, English philosopher (1907-1994).

IDEALISM: IT IS ONLY THE SPIRIT WHICH EXISTS AND DREAMS THE MATTER. Main representatives: In ancient Greece; Plato, Founder of the Academy (427-347 to B.C.) and Hegel, German philosopher (1770-1831) in Modernism. Rationalist.

**MATERIALISM:** THERE IS ONLY THE MATTER EXISTING AND EVEN EMOTIONAL EXPERIENCES AND THE SO-CALLED SPIRITUAL EXPERIENCES ARE ONLY A SUB-PRODUCT OF THE SAME MATTER AND THEREFORE THERE IS NO CONCEPT OF PERSON. Main representatives: Ludwig Feuerbach, German philosopher (1804-1872), Karl Marx, German philosopher (1818-1883), Bertrand Russell, English philosopher (1872-1970).

We must say that before the previous conceptions, **REALISM** prevails.

## WHY?

Well, in the first place, Materialism was definitively discarded by Positive Science by broadly certifying the beginning of the universe, since the fundamental thesis of this false doctrine was the supposed "eternity of matter."

It is clear then that; the "miracle" of existence, the why of the laws that govern the intimacy of matter and the entire universe, the prodigy of life, the fact of consciousness itself, as well as the sphere of values as an evident manifestation of the spirit , they are altogether something so big and mysterious that they could not be explained with an absurdity of supposedly fortuitous chemical and physical combinations.

But let an authoritative opinion illustrate this refutation. Bochenski, since the year 1947, expressed himself like this: "*Materialisms are theoretically very weak, they remain almost at a pre-Socratic level*", "*As a whole, current Philosophy has far surpassed not only its theses, but even the approach of the problems.*" Prof. JM Bochenski, Philosopher, University Professor, Mathematician and Polish writer of renown in Europe (1902-1995) ("The Current Philosophy." Bochenski, Fondo de Cultura Económica, pages 93 and 94).

IDEALISM, in stating that there is only an absolute spirit that "dreams" to be also the individual, is a clear pantheism (everything is god). Carried to the last consequences, it denies the freedom of the human being that is reduced to a moment of the necessary evolution of that one spirit.

Also, as a refutation, it is important to mention that if we bear in mind the Aristotelian sentence: "*There is nothing in the intelligence that has not passed through the senses*", we can argue; We, the real individuals, cannot "dream" anything that we have not really seen or touched. Example: If there were someone who had never, by special conditions of his birth, seen, heard, touched or felt, etc. ANYTHING, his conscience could not imagine or "dream" a reality, because lacking all experience, these functions cannot be given, which, in normal individuals, are always present faculties, although they emerge from the subconscious or even from the unconscious where they had settled.

Understanding the capital role of the senses, the conception of Realism, spirit and matter, individual intelligence and objective world is implicit.

Idealism is incapable of responding to the specific problems of man, such as good and evil that are the product of the decisions of each individual, the concept of person, etc.

Realism, as we have said, recognizes the existence of spirit and matter and synthesizes them through the concept, which intelligence forms, taking the universal from the sensitive and concrete data.
It is not only the value of common sense and that intuition that we all have regarding these two realities (*spirit and matter*) but also, when inferring conclusions provided by positive science; as a universe that has begun and that according to reason, requires a higher reality (*transcendence*) and different to it, we can conclude that in this realism is the truth.

At the height of Greek Philosophy, Aristotle (384-322 BC) the most famous disciple of Plato (although opposed to him), who developed the formal logic and gave it the character of science, gives us the great conception of first IMMOBILE MOTOR by means of an exquisite and famous reasoning; *Observing the world we see that there is BEING and MOVEMENT, but we also observe that every being is moved by another one (the being itself and the movement was given to us) and this one by a previous one and this one by another and so on as in the chain of cards of a deck that formed, one to another, they transmit an impulse, but if we continued forever like this, we would fall into the absurdity of the infinite serie, because to continue would be idle, since it would be better to affirm then that we never arrive at the origin, it is to say, that there is not a real one who gave the initial impulse, incurring in a contradiction, because the infinite serie tells us that there is no one who gave the first impulse and yet we confirm that here, there is BEING and MOVEMENT.*

In summary; if there is no one who gave that first impulse, then how is it that we, here, confirm THE BEING AND THE MOVEMENT?

The only solution is a BEING that moves and is not moved, a Being that is by itself and thus, our existence, that of the world and the movement, are understood!

Aristotle goes further, he states that:

"*This being is pure intelligence and his being is completely immaterial*"

He describes it as:

"*Thought that thinks to itself and moves without being moved*"

Later, Thomas Aquinas (1225-1274) who also represents the peak of scholastic thought, created a Philosophical, systematic and vast work called Thomism, which addresses all aspects of our reality.

He based it on Aristotle for such a colossal work. His famous 5 proofs of the existence of God are based on the reasoning of the Aristotelian Immobile Motor. He makes explicit the attributes of God from a natural rational point of view, distinguishing as the first and most important the ASEITY (*A-SE, which means by itself.*)

Aquinas proposes the proof a posteriori, that is, from sensitive facts as movement, contingency (*the characteristic of an entity to come to be and then cease to exist*), the order of the universe, the graduation of perfections, with the use of the Principle of Causality deduces that all facts demand the existence of a necessary being, the first cause, perfect and giving order to the universe, concludes by saying: "*That being is what we all call God.*"

Now let's do <u>Philosophy of Science</u>.

As we have seen, it is not enough just to be a great scientist, but we must also be philosophers, this is the key, that is, to combine both disciplines to aspire to embrace the whole of knowledge within the reach of human capacity, taking our premises from truth and scientific facts and reasoning, reaching greater conclusions, in a completely and not partially way, of our world.

## **THE CONDITIONS OF ENERGY**.

Let's recall, in traditional physics, energy is defined as *the ability to perform work or an effect*.

Let's analyze the simplest form of mechanical energy; in the transmission of an impulse.

What is required to arouse this particular energy? Well, a certain impulse. Example of the key in the machinery of a piano; First, my decision, will to produce it, in addition to the reasonableness that it will be with a purpose and also the ability or efficiency to be able to give that impulse. Now, we have that behind a certain energy there is a "will", a "why" and a "power" or "ability" to do so. In short, for the "sound-effect" of a piano key, it was necessary to have a PURPOSE that moved MY WILL accompanied by MY ABILITY or POWER to produce this impulse that was transmitted from my finger by pressing the key.

For the " universe-effect " with all its manifest and even unsuspected energy, also needed a "WHY", a "WILL" and a proportional "POWER" and then the question is "What produced it?" or should we say "Who?"

Why do we say for the " universe-effect "? because the universe is the result of a Great Explosion, which, according to the laws of Physics and Reason, is a REACTION that needed an ACTION, also coming from a WILL and with a PURPOSE, then we can say that ;

## The universe is the verifying reaction of an action, the exercise of an active will.

Given the proportions, not even really imaginable, billions of galaxies, each of them containing millions of stars, nebulae, matter and dark energy and other celestial bodies expanding indefinitely, **who could produce a proportional action for that purpose? And although that action escapes the field of experience; positive science, this action cannot be denied and it must be recognized, because without it we would not be here.**

It is then on continuity, with rigorous scientific knowledge, on a rational, metaphysical level, (*Metaphysics = said in easy words, rational science that studies immaterial objects although we know that Aristotle called it, at his time, "First Philosophy"*) continue with this study as we had already explained.

In the Einstein's Theory of Relativity, the famous formula is obtained as:

$$E = mc^2$$

Where "E" is the energy to be obtained (*in ergs*), "m" expresses the mass and "$c^2$" is the speed of light (*approximately 300,000 km per second*) squared, that is, multiplied by itself (*300,000 by 300,000*) but ...

## WHAT DOES THIS MEAN?

This means that we will obtain an enormous amount of energy only with making enter in atomic reaction <u>**ONE SINGLE GRAM OF MATTER**</u> ...

*<u>WHAT IS ATOMIC REACTION?</u>*

*The Atomic reaction is achieved <u>**by FISSION**</u>, that is, by breaking a core of Uranium or <u>**by FUSION**</u> making two Hydrogen protons unite overcoming their electrical repulsion, transforming into Helium, both produce the known chain reaction of atomic explosions, transforming matter into energy .*

*(<u>We will try to give the simplest explanation that can be done about this.</u>)*

... <u>**ONE SINGLE GRAM OF MATTER**</u> which we will multiply by the square of the speed of light, that is, 300,000 by 300,000 which gives us an amount of *900,000,000,000,000,000,000 ergs of energy*! *This is the reason why atomic bombs are so tremendously destructive.*

"AH! But arithmetic is relentless! If a <u>**single gram of matter**</u> can be converted into an amount of energy equal to that one produced by the combustion of 32 million liters of gasoline, then <u>**it will take all that energy to "make" a single gram of matter**</u>. "(*It can be verified in: "One hundred Basic Questions on Science" by Isaac Asimov, Editorial Alliance, page 129)*

Now, we will try to imagine the energy that would be needed to "manufacture" a kilo of matter, a ton of matter or the total sum of all the atoms of Planet Earth and the Solar System or of the ones in the entire Galaxy and finally of all the other millions and millions of galaxies of our universe, this would give us the **NOT EVEN IMAGINABLE REALITY OF THE ENERGY CONTAINED IN THIS UNIVERSE** !!!

All this energy was deployed in the Big Bang (*to every action corresponds a reaction of equal intensity*) **WHAT WAS THAT WHICH PRODUCED THAT BIG BANG?**
We have only one possible answer, according to the most rigorous principles of reason, if we reject it; **NOTHING WOULD HAVE SENSE, NOR THE SAME CONCLUSIONS OF THE OTHER SCIENCES**, because the monolith of knowledge would lack sustenance.

And if our cosmos, our NATURAL REALITY of which we are part initiates there, **this action unimaginable in power as the effect shows us, COULD ONLY PROCEED FROM A DIFFERENT ORDER, OTHER THAN NATURAL** (*in the impossibility that what did not exist could have been done by itself*) **THE SUPERNATURAL** !!!
That is, a form of BEING **outside space** (*then incorporeal, immaterial*) and **outside of time**, (*therefore, eternal*), without beginning or end and almighty.

# What does this look like?

Let's become aware of it, we are here, reading this book and we are part of a real universe that began at some point, as has already been scientifically demonstrated. Now that we could understand well how matter is transformed into energy in a nuclear atomic reaction and how the energy transforms in matter, *this conceived like possible in the mathematical equations of Einstein and possibly obtained some atoms in laboratory, because more than that, as already it has been explained, when not being able to take hold of so much energy as to produce greater quantities* ... does It corresponds only to God to do it in reality? But ...

Let's not be surprised! Some of the most renown scientists have testified regarding the Big Bang certification; "*We have never been closer to the "Let there be Light*". George F. Smoot, creator of the COBE satellite, the first one to certify Background Radiation; the echo of the Great Explosion that proved the beginning of the universe, expressed himself as follows: "*If you are a believer, it is like watching God*" ("*Show me God, What's the Message From Space Is Telling Us About God?*" Heeren, Day Star Publications, page 174).

Some physicists of worldwide renown, when talking about the universe, have expressed themselves as follows:

"*It seems difficult to avoid the conclusion that the current state of the universe has been 'chosen' or selected from a huge number of possible states, all of them disordered except for an infinitesimal part. And if such an initial, totally improbable state was selected, surely there had to be a selector or designer to choose it."*

**Paul Davis.**
(https://es.m.wikiquote.org)

(*There are some opinions in the sense that there would be a number of multiple universes and that the conditions for equilibrium and life have been given in ours.*
*But such a claim is based only on the imagination, <u>not on the reason</u>, because there is no real indication or observation to support such a claim.*
*The certainty and truth of science are not the product of fantasy but of demonstration.*)

The same Albert Einstein in 1929, after observing by telescope the displacement of the galaxies demonstrated by Edwin Hubble, wrote, not only of the necessity of a beginning of the universe, but of his desire to know how God created the world. "*I do not care about this or that phenomenon in the spectrum of this or that element, I want to know His Thought, the rest are details.*" (*"Show me God, What's the Message From Space Is Telling Us About God?" By Fred Heeren. Day Star Publications, Preface, Page XX*).

We must say again, do not be surprised, more and more scientists, those ones who devote their lives to fully research, have taken a complete turn in their thinking after knowing a beginning of the universe and are already <u>Deists</u> or <u>Theists</u> (*Deism = position that recognizes a personal creator God, but deny all Supernatural Revelation and the miracle, staying on a purely rational or natural level*) (*Walter Brugger Philosophy Dictionary, Herder Library, page 144*).

We do not make an exhaustive list here with the names of each of them documenting what we affirmed as we have been doing throughout this work because it is not the essential object of this work, but of course there are also great scientists around the world who confess themselves as Theists (*Theism = Posture that considers God as a personal, supramundane being, which, by its creative act, called the world from nothing to existence - creation – this posture defends the conservation of creatures by God and His continuous cooperation, His Providence and His extraordinary intervention as the Revelation and the miracle*) (*Walter Brugger Philosophy Dictionary, Herder Library, page 499*).

For example:

**Francis Collins**, one of the most prominent current geneticists, known for leading the famous Human Genome project for 9 years, named Director of the National Institute of Health by former US President Obama who considered him "*One of the best scientists in the world.*" (*Www.enghels.com*)

**Mathew Chandrankunnel**, writer, scientist, philosopher and theologian, Professor of Philosophy of Science at Dharmaram Vidya Kshetram and at Christ University, both universities in Bangalore, India. Author of several books including "*Philosophy of Quantum Mechanics*" and "*Ascending to the truth: The Physics, Philosophy and Religion of Galileo Galilei*"
(*https: //en.m.wikipedia*)

We could continue enumerating great scientists, certain of the truth of a creator God, but this is enough for the moment, because we must go ahead with the plan proposed in this work; the confrontation of Philosophical-scientific Theses with what Holy Scripture tells us in its different Texts.

# CHAPTER 3
# WHAT DOES THE BIBLE SAY?

The Bible, plural word of Greek origin meaning: *"The Books."* It is divided into two main parts; The Old and the New Testament.

**The Old Testament** is divided into The Pentateuch (*the 5 First Books*) from Creation to Deuteronomy which is the last text of the Torah (*Law given by God to his people Israel*), the Historical Books that tell the stories of the most important personages of the Hebrew people, the Sapiential Books that speak of the Wisdom of God; Its maxims in the form of Proverbs to the sublime praises expressed in The Psalms and finally The Prophetic Books that tell us the life and message that God gives to His people through 18 concrete men who are The Prophets, from Isaiah to Malachi.

**The New Testament** is composed of 27 Books. The Four Gospels (*Matthew, Mark, Luke and John*) where there is the essential of the Christian message, because they narrate Jesus Christ's facts and sayings. Then, the Acts of the Apostles written by St. Luke and later the Letters of St. Paul to the evangelized communities, the Epistle of St. James, the two of St. Peter and the 3 of St. John, the Epistle of St. Jude and finally The Apocalypse.

The Bible describes an Eternal, Creator, Almighty, Infinite God that transcends space and time, Pure Spirit who is a Justice but also Merciful Person who communicates Himself daily with His people and who describes Himself as **<u>THE BEING IN PERSON</u>** ( *Ex. 3,14 "I AM who I am"*).

## <u>The moment of confrontation has arrived</u>.

Recall, positive science speaks of a beginning of the universe and although it cannot go further, being of an empirical nature, however this assumes an ACTION that caused this REACTION that is the universe, which as we have seen, cannot be denied *(ACTION-REACTION)* but must be affirmed by reason.

What all believers would expect then from science is that one day we would discover that the universe did begin apparently from the "nothing" *("nothing" in a material sense, that is, what is subject to space and time)* and This agrees perfectly with the Bible that tells us that God created it from nothing, but not *"nothing at all"*, because God, Pure Spirit was already in the beginning (*Gen. 1,1*) He, who is the Being Himself . (Ex. 3,4)

Current science, with all certainty, tells us that the universe began in a huge flash.
The Sacred Scripture; describes it like this:
"And God said, Let there be light, and there was light" (Gen. 1-3)

## More concordances:

**The Creator Wisdom of God.**

*" From everlasting, I was firmly set, from the beginning, before the earth came into being."* (*Proverbs 8,23*)

Science of Philosophy tells us that the "*immobile Motor*" is a being by itself that originated the being and the movement, prior to everything and thus, not subject to space or time and therefore immaterial and eternal.

## Other agreement:

*" He who sits enthroned above the circle of the earth, the inhabitants of which are like grasshoppers, <u>stretches out the heavens like a cloth, spreads them out like a tent to live in</u>."*
(*Isaiah 40,22*)

We know that the Bible is not a book of positive science, but we cannot help but wonder when it is known, <u>thanks to science, that the universe is expanding</u>.

## ANOTHER:

*" I look up at your heavens, shaped by your fingers, at the moon and the stars you set firm-*
*what are human beings that you spare a thought for them, or the child of Adam that you care for him?"*
(*Psalm 8, 4-5*)

Remember, the scientists who hold Anthropic Theory say:

*"In Cosmology, the Anthropic Principle states that any valid theory about the universe must be consistent with the existence of the human being, in other words: If in the universe certain conditions for our existence must be verified, these conditions are verified, since we exist"*
(*Anthropic Principle, en.m.wikipwdia.org*).

Agreeing with those who claim that the universe was created especially for humanity.

## ONE MORE:

*" Can you fasten the harness of the Pleiades, or untie Orion's bands?*
*Have you grasped the celestial laws? Could you make their writ run on the earth?*
*Will lightning flashes come at your command and answer, 'Here we are'?"*
(*Questions from God to Job, Jb 38: 31,33 and 35*)

Remember that when we talk about all the energy necessary to create the universe, we discover that only an Almighty Being could have created it; given the proportions. A Being who also maintains His Lordship over His creation.

The following Biblical citations illuminate some scientific-philosophical problems that exceed our capacity to respond, for example:

We all have a specific location in some town or city, for example; the City of Guadalajara, which is in the country Mexico, which is located in the American Continent and this also, on the Planet Earth and the Earth itself is part of the Solar System and the Solar System it is part of the Milky Way; our galaxy and this, together with the Andromeda galaxy, the Triangle Galaxy and some other 30 smaller galaxies more, form the Local Group, which is contained within the Virgo Cluster and this cluster, along with millions of other galaxies in other clusters they make up the expansion (*as it is known, the SPACE-TIME goes "doing" with the universal expansion*) And then ...

## WHERE IS THE UNIVERSE EXPANDING ??

Apparently there is no answer, but when scrutinizing in the Sacred Texts; and given everything explained in this work, we find a Text that pleases the understanding in a wise and simple way; Saint Paul tells us:

"*For in Him (God) we live and move and have our being.*" (*Acts 17,28*)

Continuing with the concordances, in Chapter 1, page 10 of this work when we speak of the Cold Death of the universe and the total degradation of its energy, as a consequence of an expansion that will occur indefinitely until its end, since the universe is <u>an open system</u>, we also find the following Biblical Texts:

*" Lift up your eyes to the heavens,
    look at the earth beneath;
the heavens will vanish like smoke,
    the earth will wear out like a garment
    and its inhabitants die like flies."* (Isaiah 51,6)

## And also:

*"Heaven and Earth will pass away, but My words will never pass away"* (Matthew 24,35)

## AND:

*"See, I will create new heavens and a new earth.
The former things will not be remembered,
nor will they come to mind."* (Isaiah 65,17)

## One more:

*"Since ancient you founded the Earth
And the Heavens are the work of Your Hands,
They perish but You stay,
All of them like clothes wear out,
Like a dress you change them
And they move, but You are always the same,
Your times have no end. "*     (Psalm 104, 2 and ss)

*(We insist, we do not look for scientific facts in the Bible, but in the case of the ultimate truths, we cannot fail to point out their conformity)*

In Chapter 2, page 25 of this book, when we talk about the ACTION that produced THE REACTION THAT IS THE UNIVERSE, we clarify that positive science cannot reach that ACTION, by escaping from the field of experience, that is, although that ACTION is true, we will never "see" it.

There is another wonderful concordance expressed by Saint Paul in his Letter to the Hebrews:

*"It is by faith that we understand that the ages were created by a word from God, so that from the invisible the visible world came to be."* (*Hebrews 11,3*)

Let's continue, in Chapter 1, page 5, it is mentioned that everything started in a so-called Big Bang or Great Explosion where they started, in unison, the **ENERGY-MATTER and the SPACE-TIME**. We cannot fail to mention St. Augustine (*Christian Philosopher and Theologian 354-430*) one of the Holy Fathers, (*Holy Fathers: First commentators of the Holy Scriptures who interpret and explain them, placing them within the reach of the people. Tradition, one of the three elements of Revelation.*)

Surprisingly, since the fourth century, St. Augustine tells us:

*"That the beginning of the creation of the world and the beginning of time is one and that it is not one before the other"* *"because there could not have been a time in the past before the world"* (*The City of God, San Agustín, Porrúa Editorial*), *Pages 291 and 292*)

Bear in mind that with the Theory of Relativity we assume the <u>SPACE-TIME as an indissoluble unit</u>, so that it is better understood, the space is something like the "form" of time, as they have already said.

## We also find in <u>Verse 2 of Genesis</u>:

*" the earth (understand not our planet, but what would later be the matter) was a formless void, there was darkness over the deep, with a divine wind sweeping over the waters."*

In Astrophysics, specialists talk about a <u>singularity</u> in the first moment, which means that the laws that govern the universe today; Gravity, Electromagnetism, Strong and Weak Atomic Forces, as we know them, they still did not operate at first.

## Another agreement:

The Thomistic Philosophy tells us that to prove the existence of God, we must infer in a posteriori demonstration (cause and effect) from something concrete, going back to the ultimate cause that is a necessary Being. ("Necessary" in philosophy means: INAMOVIBLEMENT TRUE, WITHOUT WHOM NOT)

And the Letter of Saint Paul to the Romans in Chapter 1,20 tells us:

*"Ever since the creation of the world, the invisible existence of God and his everlasting power have been clearly seen by the mind's understanding of created things. "*

And in the Book of Wisdom, in the first part of Chapter 13, especially in Verse 5, you can read:

*" Since through the grandeur and beauty of the creatures we may, by analogy, contemplate their Author."*


After having seen the previous concordances, it is also convenient to mention, for Catholics, that former Pope Benedict XVI affirmed: *"God was behind the Big Bang,"* he said; *"The universe did not arise by accident"* all this was sustained by the Pope on the Day of Epiphany, Friday January 7, 2011, in a sermon, in St. Peter's Square, before more than 10,000 people, in addition to affirming that *God could also use a process of natural evolution.* ("*El Universal" newspaper, Friday, January 7, 2011*).

The current science in its vertiginous progress, more and more, **discovers a world that reveals a directed organization always seeking a purpose,** not only in the question of the beginning of the universe, but also and especially in Biochemistry and Genetics, such situation that fits perfectly with the action of a God Creator and Organizer of His universe.
That is why we intuit that the very existence of human beings has, also as one of its purposes, the sharing of the

greatness of His works through this temporal window that is our life.

The, in another time, distance between Faith and Science, after the portentous achievements of science, is dramatically reduced and if we use, in addition, the reason doing - Philosophy of Science - this new discipline, (*taking as premises the experimental data*), we will have found the means of their agreement; clear, always each, Science and Religion, in their respective planes, language and identities.

## SUMMARY:

We offer a very brief extract of what is contained in this book for readers who are not well versed in the topics discussed here, so that they can more easily retain the fundamental reasons for what has been concluded here.

## **THE ORIGIN OF THE UNIVERSE**.

Because of Science we know that the universe has started in a Big Bang.
The great astronomers: George Lamaitre (*1920*) and Edwin Hubble (*1929*) and the Astrophysical scientist George Gamow (*1946*) had theoretically explained, based on their observations and calculations; the origin of the universe in a Great Explosion. Gamow had already developed, formally, a physico-mathematical theory, which

predicted that the expansion would continue indefinitely. (*It is now known due to the most recent observations and the current calculations on critical density (2017) corroborate this. Https://ciencia.nasa.gov/science-at-nasa/2001/ast03apr_1*) The discovery of Dark Energy in 1998, which is 73% of the universe!!! Tells us today that any hypothesis of a contraction is ruled out. (*It is necessary to update ourselves at this point and that the reader, in case of wanting to verify this data, does so in sources after 1998, preferably from 2010 onwards, where this is already clearly concluded*).

Previously, the Russian mathematician, specialized in Relativistic Cosmology, **Alexander Friedman**, in the **1920s** had warned **Albert Einstein** that his equations showed an expanding universe, this was not accepted by Einstein immediately, because he conceived a static universe and without beginning, but when **Edwin Hubble**, in 1929, in his observatory, shows, at Einstein's own eyes, the estrangement of the galaxies in their redshift (*deduced this from the Doppler Effect*), Einstein renounces his belief in an eternal universe and admits that this must have had a beginning.

The Great Theory then took a rigorous form, but that is not all. Gamow predicted that if this was indeed the case, the "echo" of the Big Bang should still remain in space.

By mere accident, in 1967, two scientists from the Bell Company: **Arno Penzias and Robert Wilson** discovered this radiation while they were working on the use of antennas to perfect satellite communication.

The Physicist **Robert Dicke**, who also sought on his own to find this radiation, is the first one to identify it in the achievement of the forementioned Penzias and Wilson to whom they were awarded the Nobel Prize in Physics for *"Their extraordinary discovery."*

As if something were missing, NASA (*the National Aeronautics and Space Administration in the USA, the maximum scientific and technological center today*) built and launched to space in 1989 the COBE satellite; an artifact with refined measurement equipment and later the satellites WMAP (*2001*) and PLANCK (*of the European Space Agency 2009*) for its definitive verification and they fulfill their mission verifying it; indeed, microwave radiation from all corners of space, as expected according to predictions, is the fossil rest of the Great Explosion that originated the universe and even the beginning of time (*together* with the expansion of matter , at once it is being created the space-time).

After meticulous studies and analysis, the news is formally disclosed in all specialized science journals.

We can conclude, according to the data of positive science, that our universe started from a Great Explosion-Big Bang. **Where was the alleged "eternity" of the matter so preached in the nineteenth century and early twentieth?** Any doctrine of materialistic type is then refuted, which we already knew for the right reason. <u>There is no greater contradiction than a conscience denying itself in materialism</u>.

The principle of The Conservation of Matter, in which some people believed to see the eternity of it, refers to matter understood as weight and is born from the experiments of Lavoisier in 1772. He found that in all reactions, including combustion, *the weight of the compounds is equal to the weight of the components before and after the reaction*, but time would let us know
that things were not as simple as they were then believed, because the most surprising of the reactions was unknown; The nuclear reaction. In this, under the concept of matter understood as weight or mass, it is impossible to understand the previously concluded, as already explained in detail above, due to the so-called "Mass Defect"; that little radiant energy that is lost and leaks in the form of light when atomic fusion takes place.

This light, like that of the Sun (*which is an atomic furnace like all stars*), wanders through all space and if it is absorbed by an opaque body it is transformed into heat and performs work, movement and / or cools, thus the concentration of the energy is disappearing very slowly but necessarily, this because the universe ...
EXPANDS!!! Energy and heat "is not enough" to keep it stable.

The Big Bang shows that <u>THE UNIVERSE ITSELF IS THE VERIFYING REACTION OF AN ACTION (*action-reaction*) THE EXERCISE OF AN ACTIVE WILL,</u> and although, with positive science we cannot study that action in itself, because it transcends our reality, we cannot deny it and we must affirm it rationally, because without it ... **we would not be here!**

----- Science tells us the universe; matter-space-time, began.

----- It is evident that out of nothing; nothing comes, so, the universe comes from a Being beyond matter, space and time. Of course, this action was carried out by a transcendent BEING of a different nature.

----- The Bible reveals to us a Spiritual, Transcendent and Almighty God who created the universe.

In addition to the two levels; Revelation and Reason that have always discovered a creator ("*Ever since the creation of the world, the invisible existence of God and his everlasting power have been clearly seen by the mind's understanding of created things.*" *Rom 1.20* ) we also find now that positive science points with rigorous precision towards the same fact.

Echoing the great Professor Manuel García Morente (*Preliminary Lessons in Philosophy, García Morente, Pages 293 and 294, Editorial Purrúa*), saying that a new era of thought begins when he quotes Don José Ortega y Gasset; we can conclude with the context of the words of this one; *The prow of the ships has had full turn and must navigate, as never before in the field of knowledge, towards the horizon of the Divinity;* "**GOD AT SIGHT!**"

We recognize Reason as the immovable column where right thinking is based, which reveals to us a "*Giver of movement and origin to the universe*" and we value the confirmations that positive science gives us in its language, as we have seen, of these facts.

We conclude this investigation with the same title of the present book; **"Was the universe truly created by God?"** But now in an affirmative way:

## YES, THE UNIVERSE WAS TRULY CREATED BY GOD !!

ALFONSO CASTILLO G.

# **THE TRANSCENDENT TRUTH.**

I searched inside, questioning myself
And the self in question is so fragile
that it vanishes without scent,

For there is only one sustenance of existence and form
And energy and matter are held to His Norms.

Everything encompasses this principle; Wisdom and Good,
Its flow is called Life, it governs the inert, too.

(1) - In the ecstasy of Love or behind the darkest night,
He can be felt--; find Him is a sweetness-bright!

(2) --- More intimate to us than ourselves --- as it was said in a temple,
His "obsession" is to share; the Creation is an example.


It is never early or late in the destiny of a being,
For events and time are one with His Will.

    And why to fear death with pain and sadness?
(3) --- If we are and move within His Thought and Loveliness---

    It is impossible to doubt in all this conception,
(4) --- Because His Works, gives Him accreditation---

(5) --- This Being moves everything --- even cosmos comes from Him,
As perfect the principle is; (6)--- Nothing comes from nothing, that's the quid-

Since to man, reason requires perfect judgments;
-God is the cause of things, beings are His effects-

Philosophy and Science met with Faith
And in his evolution man will return to believe.


*(1) Saint John of The Cross.*
*(2) Saint Augustine, Bishop of Hippo.*
*(3) Saint Paul.*
*(4) San Paul.*
*(5) Aristotle.*
*(6) Universal Logical Principle.*

| SUPERNATURAL PLANE | OUR TEMPORAL-SPACE REALITY |
|---|---|
| "LET THERE BE LIGHT"<br>ETERNAL<br>**ALMIGHTY** | THE CREATION.<br>(ACCORDING TO THE BIBLE) |
| **GOD** – IMMOBILE MOTOR | ORIGIN OF THE BEING AND MOVEMENT |
| PURE SPIRIT   CAUSE | EFFECT<br>(ACCORDING TO PHILOSOPHY)<br>REASON |
| TRANSCENDENT | |
| THE GREAT ACTION'S AUTHOR | VERIFIED REACTION OF AN ACTION.<br>BEGINNING OF SPACE-TIME<br>13,600 MILLONS OF YEARS<br>(POSITIVE SCIENCE SCOPE) |
| **SUPERNATURAL PLANE** | **OUR TEMPORAL-SPACE REALITY** |

Finally I conclude this book that corresponds to the commitment I acquired almost 30 years ago; to honestly recognize the truth, that one I would find after the investigations, whatever it were and now I express it in this reflection entitled  "WHO?"

# *WHO?*

*THE GREATEST INTUITION THAT ONE CAN BE GIVEN;*
*ARRIVE AT ASKING;*

*HOW IS THE BEING POSSIBLE? WHO PROMOTES IT?*
*WHO WITH THE SPACE AND TIME CAN DO IT?*

*WHO, PREVIOUS EVERYTHING, HAS ALWAYS BEEN?*
*WHO IS SO OPPOSED TO NOTHING? FEELING MOVED*
*I ASKED!*

*WHO COULD DEFINE THE LIGHT AND THE EXISTING LAWS?*
*WHO SOWED SO MANY SPECIES AND MADE THEM DIFFERENT?*

*AND WHY I, ON "AWAKENING",*
*MEET EVERYTHING FREELY THERE?*

*AND TO WHOM I PRESENT SO INSIDE ME, KNOWING EVERYTHING*
*THAT IN MY CONSCIOUSNESS LIVES IN SOME WAY,*
*THUS WITH INSISTENCE I HEAR HIS VOICE?*

*ONE IS THE ONLY ANSWER:*
## *GOD!*

## BIBLIOGRAPHY.

We recommend these works for the reader who wants to deepen the topics discussed here.

---- "Introduction to the Scientific Method". Raúl Gutiérrez Sáenz. Esfinge Editorial.
---- "The Meaning of Relativity". Masterpieces of Contemporary Thought. Albert Einstein. Planet Editorial.
---- "One Hundred Basic Questions on Science". Isaac Asimov. Alliance Editorial.
---- "Relativity for Beginners". Shahen Hacyan Science from Mexico # 78. Fondo de Cultura Económica Editorial.
---- "Space, Time and Gravitation" (The Big Bang Theory and Black Holes) Robert M. Wald. Fondo de Cultura Económica Editorial.
---- "A Universe in Expansion". Luis F. Rodríguez. SEP Fondo de Cultura Económica Editorial.
---- Encyclopedia of Astronomy and Space. "The Universe" Planeta De Angostini Editorial.
---- "Show Me God. What The Message From Space Is Telling Us About God. " Freed Heeren Day Star Publications.
---- "Stars, Clusters and Galaxies". Martín M. Rees. Salvat Publishers Library.
---- " Quasars in the Confines of the Universe". Déborah Dultzin. Science for all. Fondo de Cultura Económica Editorial.
---- Magazine Collection "Science and Technology". CONACYT.
---- "Science, Reason and Faith". Mariano Artigas EUNSA
---- "Philosophy of Science". Mariano Artigas EUNSA

---- "History of philosophical Doctrines". Raúl Gutiérrez Sáenz. Esfinge Editorial.
---- "Introduction to Logic" Raul Gutiérrez Saenz. Esfinge Editorial.
---- "Logic and Cosmology". Regis Jolivet. Carlos Lohlé Editions.
---- "Preliminary Lessons of Philosophy". Manuel García Morente Porrúa Editorial.
---- "The Thought of St. Thomas." F. C. Copleston. Fondo de Cultura Económica Editorial.
---- "The Current Philosophy". I. M. Bochenski. Fondo de Cultura económica Editorial.
---- "Dictionary of Philosophy" Walter Brugger. Herder Library.
---- "The City of God" St. Augustine. Porrúa Editorial.
--- Bible of Jerusalem. Spanish Descleé de Brouwer, S.A. Editorial.

## ABOUT THE AUTHOR.

Professor Alfonso Castillo G. is a prestigious Professor of Philosophy and Sciences who has taught the subjects of Philosophy of Science and Philosophy of History for more than 13 years.
Besides being a philosopher, he is also a talented musician, pianist and composer.
In this book he pours his knowledge, Astronomy, Physics and Philosophy and confronts them with the Biblical writings seeking to find the true origin of our reality.

Comments to the e-mail:
alfonsoacastillog@hotmail.es

www.ingramcontent.com/pod-product-compliance
Lightning Source LLC
Chambersburg PA
CBHW030035230526
45472CB00002B/528